I0425309

CHINESE RABBIT NEW YEAR

DUQING GONG

HUNAN ELECTRONIC
AND
AUDIO-VISUAL PUBLISHING HOUSE
EUROPE PUBLISHING & PRINTING SERVICE LIMITED

authorHOUSE®

AuthorHouse™ UK Ltd.
500 Avebury Boulevard
Central Milton Keynes, MK9 2BE
www.authorhouse.co.uk
Phone: 08001974150

Hunan Electronic and Audio-Visual Publishing House
3E, Ying Pan Road 3, Changsha, 410005, China
http://www.xyin.com/
Tel: +86 (0)731 84456374

© 2011 Duqing Gong. All rights reserved.

No part of this book may be reproduced, stored in a retrieval system, or
transmitted by any means without the written permission of the author.

First published by AuthorHouse 6/14/2011

ISBN: 978-1-4567-7979-5 (sc)

Any people depicted in stock imagery provided by Thinkstock are models,
and such images are being used for illustrative purposes only.
Certain stock imagery © Thinkstock.

This book is printed on acid-free paper.

Because of the dynamic nature of the Internet, any web addresses or links contained in
this book may have changed since publication and may no longer be valid. The views
expressed in this work are solely those of the author and do not necessarily reflect the
views of the publisher, and the publisher hereby disclaims any responsibility for them.

Table of Contents

China today is known as "The People's Republic of China". It is an ancient and mysterious country with over 5000 years of civilization history. Located in the eastern part of Asia and on the west coast of the Pacific Ocean, it covers a general area of 9.6 million kilometers, with a population of over 1.3 billion. Today, China is also rated as one of the fastest growing countries, in perspectives of economic development and comprehensive national strength.

Just like Christmas in many countries, the Chinese Spring Festival is also a yearly event, starting from January 1st of the Lunar Calendar (Chinese Traditional Calendar). January 1st is also regarded as the first day of the new year. And every Chinese year is represented by a zodiac animal. For instance, 2008 is the year of the Rat, 2009 is the year of the Ox, 2010 is the year of the Tiger, 2011 is the year of the Hare and 2012 is the year of the Dragon...

1. An Introduction to the Chinese New Year and the 12 Animals of the Zodiac

A. The Chinese New Year

As the end of the last and the beginning of the next, Spring Festival, also named as the Great Year, is regarded to be the most important traditional festival of China. The original Spring Festival started on December 8th (The day of December Worship on the Chinese Lunar Calendar. The ancient people used to go hunting on this day and worship the ancestors with the captured prey) until January 15th. And the New Years Eve and January 1st are the most important days of the entire festival. Through over 4000 years of development, a colorful variety of folk customs have been formed for Spring Festival, including December 8th Celebration, Kitchen God Worshipping, Sweeping, Purchasing of New Year Goods, Bath Taking, House Decoration, The Reunion, Staying Overnight, New Year Lucky Money, Fire Crackers, God and Ancestor Worshipping, New Year Visits, Lantern Festival, Sticky Rice Balls and so on. All these traditional Chinese customs and cultural events are carrying the profound soul of Chinese culture and national spirits.

a. Origin of Spring Festival and difference the between Lunar and Solar Calendars

Spring Festival is also the "Year" in Chinese folklore. Celebrating the Spring Festival is just celebrating the year. In an old Chinese saying, 'New Year is the most important day among other festivals, because as the first festival throughout the year, this day should be celebrated as the most fascinating, the most amazing and the most colorful carnival'.

The creation of "Year" in Chinese history was directly related the forming of the lunar calendar. And "Year" was originally referred as the day of harvest. The growth of crops is greatly influenced by the periods of different seasons and the movements of the sun and the moon. Through years and generations of production activities, the ancient Chinese learned to follow the laws of the day and the night, the changing of the seasons and the orders of nature. And such findings were used in the

course of agricultural productions and finally drawn as the lunar calendar of the year, the month and the day. As people only harvested the crops once a year, "Year" was then used to describe the maturity of crops by the ancient Chinese. The forming of "Year" is regarded as one of the most significant events in Chinese history. People's lives, time, planting and harvesting of crops, turning of seasons and changing of temperature cycle all terminate with the year. After the good harvest, Chinese people would celebrate the success, worship their gods, give thanks to nature and pray for the next year. Even after a crop failure or a series of disasters, Chinese people would also pray for the forgiveness of the gods and the turning of the next year. Generally speaking, the development of the concept of a year was originated from the ancient human knowledge about seasons, astrometry and calendar studies.

The lunar calendar of ancient China was identified through daily agricultural production activities according to the moving patterns of the moon. With the sun as the reference object, the complete moving period of the moon around the earth (the moon has actually moved for more than 360 degrees) was called a month and regarded as the basis of lunar month. And a whole year was divided into 12 lunar months. This method was named as the Lunar Calendar. As this calendar was so important to agricultural production, it was also called as the "Agricultural Calendar". Spring Festival was calculated according to the agricultural calendar and regarded as the beginning of the year, namely the traditional Chinese New Year.

The timing of the New Year was finally fixed in Han Dynasty (Hanwu Emperor), Year 156 BC to Year 87 BC. And the first day in each lunar year was identified as the beginning of the year. From then on, different festivals were also set accordingly. On January 1st 1912, Mr. Sun Yat-sen introduced the western calendar to China, which was recognized as the public calendar and put into use. In 1914, the January 1st of Lunar Calendar, the New Year of Agricultural Calendar, was officially named as the Spring Festival. And the timing of Spring Festival was finally fixed.

In time order, in the last month of the last year and first month of the next, after months of hard work, Chinese people will use these two months to celebrate the most important festival, to enjoy happiness, feasts, worshipping and the blessing. Ideology, morality, ethics, love, aesthetics and religion are fully shown on the stage of Spring Festival. Therefore, to Chinese, Spring Festival is such an important time of the year.

b. Allusions of the Spring Festival

There was an interesting allusion about the "Year". Long, long ago, the "Year" was a dangerous and crude monster with sharp antennae, ashen face and long fangs. It lived at the bottom of the sea and always crawled to the bank at New Year Eve, destroying crop fields, swallowing farming animals and hurting human beings. It caused great disasters to the people living by the bank. People had to move their families to the mountains and hide from the "Year".

Until one New Year Eve, people saw an old man walking towards the village with ease, while people were still busy packing everything, sealing windows and doors to move their families away to the mountains. This man with a gray beard and sparkling eyes walked with a peach wood stick, carrying a bag on his arm. His behavior looked incredible and even crazy to the people living in this place. A nice old lady living in the east part of the village stopped him by grabbing his sleeves. She didn't want him to get hurt by the "Year" and asked him to go with everyone else. Unexpectedly, the old man stroked his beard, smiled and said: Don't' worry Ma'am. Could you let me borrow your house just for one night? I will drive the "Year" away. Hearing this, the old lady started to believe that this old man was crazy. She then left with the other families. In the mid night, the "Year" entered the village and started looking for animals and human beings. However, the village was already empty. When it was about to leave disappointedly, the smell of fresh meat from the east part of the village stopped its pace. It then traced the smell back to the house of the old lady and found a red paper pasted on the door. The entire house was lit up and a pile of burning bamboo was cracking in the front yard. With scares, the "Year" hesitated for a moment. But it just could not resist the attraction of the smell. In a scary scream, it ran to the door. Right then it heard the sounds of fire crackers from the yard. The monster's body was shaking with fears and it never dared to get closer, not even a single step. The monster turned out to be afraid of the color red, fire and the sound the fire crackers. It suddenly saw the door of the old lady open. An old man in a red robe walked out with a peach wood stick in his hands. He laughed loudly and walked towards the "Year". The monster was so scared it screamed and ran away as fast as it could.

On the next day, when the people came back and found that the entire village had remained unchanged, everyone was surprised. And the old lady suddenly realized that all of this was because of the old man. People rushed to the house of the old lady and found the red paper on the door, the cracking bamboo in the yard and the red candles in the house. The old man walked out of the yard and taught people the tricks drive the "Year" away. This news was spread quickly to other villages. From then on, in every New Year Eve, every family started to hang red couplets, to burn firecrackers and to light the houses up with red candles. And this also became an annual custom and carried on by generations of people.

B. Introduction to the 12 Zodiac Animals and the Hare

a. Introduction and Allusions of the 12 Chinese Zodiac Animals

Introduction of the 12 Zodiac Animals

Different from the Western Zodiacs, like Taurus, Leo, Sagittarius and etc., Chinese people believed in 12 Animals, such as the Ox, Tiger and so on, namely the 12 Zodiac Animals.

The 12 Zodiac Animals in Chinese culture are composed of 11 real animals, including the Rat, Ox, Tiger, Hare, Snake, Horse, Ram, Monkey, Rooster, Dog and Pig and 1 legendary animal, the Dragon. It is not only an interesting way to calculate the year, but also an attractive part of the culture. In Chinese people's hearts, these animals have become the symbols of certain supernatural powers. The turn of the 12 years starts from the Rat and ends at the Pig. Then the cycle will start again. Every Chinese person has a Zodiac animal for the year that they were born in. People also believe that different Zodiac animals represent different destinies and different endings of their lives. It is also related to the personalities of the people. Different Zodiac animals will gift different personalities to the owners. Anyone who's interested to find out about his own Zodiac animal and personality analysis, can check the query chart in the appendix.

So how did people pick the 12 Zodiac animals? In the primitive society, with extremely low productivity and limited knowledge about nature, the lives of Chinese ancestors greatly relied on different kinds of animals, such as horses, rams, oxen, roosters, dogs and so on. And people were also in fear of some dangerous animals like tigers and snakes. And some animals with superior natural capability to human beings were also respected by people, such dogs with their sharp sense of smell and monkeys with super agility. As a result, people created and started to believe and worship the supernatural ability of the 12 Zodiac Animals. People then put the figures of 11 real animals and the dragon together with time and primitive beliefs, to calculate the year, the month and the time. The same as western zodiacs, there are also 12 animals to represent the time in China, because in ancient China, 12 is a very important number, which was widely appreciated by people. As a traditional custom, a Turn includes 12 years and a year includes 12 months. And ancient music includes 12 tunes. It is a key number for both the ancient calendar law and folk customs, as well as a key number in Chinese culture. That's also why people picked 12 animals to represent the years.

The 12 Zodiac Animals in China not only are the symbols of the years, but also carry the expectations and dreams of Chinese ancestors to the later generations. The 12 animals were divided into 6 groups, 2 in each, to represent different meanings.

Group 1, Rat and Ox. Rat represents wisdom while Ox represents industry. The two points can never be separated. A smart but lazy guy is stupid and can't make big successes, while an industrious but silly guy is always a sad loser. The combination of wisdom with industry represents the most important expectation of Chinese ancestors.

Group 2, Tiger and Hare. Tiger represents bravery while Hare represents prudence. The perfect goal is always achieved with perfect balance between bravery and prudence. The lack of any of these, can only make an idiot or a coward.

Group 3, Dragon and Snake. Dragon represents toughness while the snake represents pliability. Either too much toughness or too much pliability will cause irreversible breaking or loss of assertiveness. Therefore the perfect combination of the two characteristics has long been recognized through the Chinese history.

Group 4, Horse and Ram. Horse represents assertiveness while ram represents harmony. Over-assertiveness, without consideration of the challenges, may be frequently disturbed by difficulties and end up missing the real goals. And over-consideration put on the environments and difficulties, may cause the loss of directions. So, a perfect man should be assertive with harmonious personality.

Group 5, Monkey and Rooster. Monkey represents flexibility, while Rooster, who crows regularly and timely everyday, represents consistency. Flexibility and consistency should be perfectly balanced, otherwise nothing will last long or be achieved with merely flexibility and zero constancy, or no innovation or development could be made with just constancy and no flexibility. Great success is usually made by both the two factors.

Group 6, Dog and Pig. Dog represents loyalty while Pig represents agreeableness. A loyal person with low agreeableness is a bad team player. And on the other hand, an over-agreeable person without loyal and belief is usually an unreliable guy with no discipline. The perfect match between the two sides is regarded as the traditional guidance of behavior, namely "Like a square in a circle, a real gentlemen seeks harmony but not uniformity"

Interesting allusions and idioms about the 12 Zodiac Animals

You may wonder why these 12 animals were picked as the Zodiacs. Here's the story. The Jade Emperor was trying to select 12 animals to be his royal guards in the heavenly palace. He then sent an immortal to announce this news to all the animals in the world. The candidates should go the palace at the set time. The earliest to arrive would be picked as winners, while the late ones would lose the game or just be ranked lower than others. At that time, the Cat and the Rat were still good friends and the Cat wanted to become the royal guard so badly. However, he had the bad habit of oversleeping. He then asked the Rat to wake him up on time. But the Rat forgot to do so. He went to find the Ox and asked for a pick, as the Ox always woke up early and ran fast. So the Ox readily agreed on that. At that time, the Dragon wore no horns, but the Rooster did. The Dragon then told the Rooster that he was already so beautiful and didn't need the horns. And the Dragon asked the Rooster to lend them to him. Flattered by the Dragon's words, the Rooster was so happy and agreed to lend him the horns. He also asked the Dragon to return the horns after the selection. the Dragon promised he would. When the time came, all the animals ran to the heavenly palace as fast as they could, while the Cat was still soundly sleeping. Riding on the back of the Ox, the Rat reached the palace. But before the Ox crossed the line, he jumped over and became the first one to cross the line. As he was the earliest winner, the Jade Emperor picked him as the first royal guard and the Ox was the second. the Tiger was the third and the Hare was the fourth. Though the Dragon was late, the Jade Emperor was so impressed by the appearance of him and then selected him as the fifth. And he also offered to rank the son of the Dragon as the sixth. However, the Dragon disappointedly realized that his son was not coming today. Right at that moment, the Snake ran over the line and said: "I'm the adopted son of Dragon and I should be the sixth". He made it to become the sixth. When the Horse and the Ram arrived, both of them were nice and willing to help the other to cross the line on time. The Jade Emperor was moved seeing this and finally ranked them as the seventh and the eighth. the Monkey was behind the thirtieth, until the last second. He quickly jumped over the cloud and crossed the line to become the ninth. The last three were the Rooster, the Dog and the Pig. After the competition, the 12 animals were ranked as Rat, Ox, Tiger, Hare, Dragon, Snake, Horse, Ram, Monkey, Rooster, Dog and Pig. When the Cat woke up and found that Rat became the first royal guard, he was so mad and started to chase Rat to every corner of the world. When the Dragon returned to the sea, he found himself so beautiful with the horns in the reflection of the water. He decided no to return the horns to the Rooster. From then on, he hid into the deep sea and disappeared from the world. Cheated by the Dragon, the Rooster was mad. From that day, he started to crow in front of the sea "Wo, Wo---Wo, Dragon, where to hell did you go. Give my horns back to me!"

As colorful folk culture of China, the 12 Zodiac Animals have been named in many idioms. Below listed are some of them.

Rat: **Shorted-sighted as a rat:** It means somebody doesn't have long-term strategy and only cares about unimportant things.

Hit the rat or hit the art: It means in fear of hitting art wares hesitating to throw things at the rat. It figures the hesitation and fear to do anything.

Ox: **Give a try with the Ox Chopper:** Ox Chopper means the chopper to kill ox. Give a try means just using the tool for unimportant job. It figures a knowledgeable person unveiling his talents from unimportant jobs. It also means that the talents of genius will be revealed on the first day of work.

Playing harp to the ox: it means playing classical harp music to the ox, while the ox knowing nothing about what happens. It figures talking to the wrong person or explaining a profound theory to the idiot. Irony intended on ignorance.

Tiger: **Live Dragon and Tiger:** It means energetic and brave like live dragon and tiger.

Tiger with wings: By tiger with wings, it means the great man is turned stronger with powerful and scary help.

Hare: **Fox cries the death of hare.** By fox crying at the death of hare, it means the sadness for the allies.

Falcon dives at the jumping hare. Falcon is an eagle like bird, raised for hunting. When a hare jumps, the falcon will dive to catch it. It means the swiftness of actions and the vigorous calligraphy.

Dragon: **Crouching Tiger Hidden Dragon:** it means uncovered and outstanding talents.

Finishing Touch on a Dragon Picture: the story tells the tale of a famous painter in Liang Dynasty, who drew four dragons on his home wall without eyes. He told people that once the eyes were painted, the dragons would fly away. But people didn't believe him and insisted that he should paint the eyes. Finally he was convinced to paint the eyes for two dragons. Right after he finished it, flashes and thunders occurred in the sky, and these two dragons flew away immediately, while the other two blind ones still remained on the wall. This idiom is used to figure the most important and impressive part of the artwork.

Snake: **Beating the grass and flushing out the snake:** the snake is flushed out by the beating on the grass. It means the stupid and unprepared actions.

Snake-like reflection of a bow in the cup: the reflection of a bow is imagined as a snake. It means the people who are too suspicious illusory about everything.

Horse: **Ride on the first horse:** it means riding on the first horse in the war. It means the brave pioneer with no fear about the difficulty.

Heavenly horse flying in the sky: the heavenly horse runs at an incredible speed, just like flying in the sky. It means the creativeness flowing uninhibited.

Ram: **Mending the fold after the rams have been stolen:** it means that repairing the fold after the ram is stolen by wolves through the holes in the fold is still not so late. It means that it's never too late to correct an error and it at least prevents the further loss.

A narrow winding trail like ram intestine: it means the winding and narrow paths and latterly figures the narrow-minded people.

Monkey: **Monkey Cheek and Monkey Mouth:** Cheek means face. Monkey mouth with big cheek figures an ugly looking.

Kill the chicken to scare the monkey: it means to scare the monkey by killing the chicken. It means the way to warn somebody by punishing other people.

Rooster: **Rooster feathers and garlic skins:** it means the tiny and unimportant things.

Small Belly with Rooster Intestine: it means narrow and stupid minds without strategic thinking.

Dog: **Be a bully dog with the backing of a powerful person:** it means the people who bully the weak ones, backed by a powerful person.

A desperate dog can jump over the wall: it means that a cornered person can do anything extreme.

Lead a dog's life: it means lay people with terrible behavior

Pig: **People hate big name and pig hates big fat:** it means people will get in trouble when he gets famous, while the fattest pigs will be first killed.

Introduction of Chinese Zodiacal Year

Zodiacal Year is a very important part of the Chinese Zodiac Animal customs, which puts the life of human beings into the reincarnation every 12 years. The years of 12, 24, 36, 48, 60 and so on are the starting ages of the next turn and reincarnation. And all these years are called Zodiacal Years. Zodiacal Years were created according to the rotation of the 12 Zodiac Animals. In Zodiacal Years, people will more easily experience unsatisfactory, ridiculous or weird issues. Therefore, Zodiacal Year has been considered unlucky for Chinese people. In another word, the lucky people will enjoy extremely good luck for a whole year, while the unlucky ones may even want to kill themselves facing bad luck for the entire year. As a result and as a custom, people worry about their luck in such years and will try all the ways to avoid it. In China, red is a color that can protect people from evil things. So in Zodiacal Years, people are used to wearing red vests, red under-wear, red socks or red belts. And they also like to wear garment ornaments on red strings, to have good luck and to pass the Zodiacal Year safely. And this has also become a great opportunity for businessmen, who have put huge resources to develop commodities about Zodiacal Years and therefore made it into a big industry.

For Zodiacal Years, people would rather believe in the existence of bad luck and wear red items just to keep bad luck away. You can see that the culture of Zodiacal Year has been blended into the daily life of the Chinese. And it's also influencing the way of life and working of modern China to a great extent.

b. Meanings and Allusions of the Hare in China

① Meanings of the Hare in China

Hare, ranked as the fourth animal in the order of Zodiac Animals, is considered as a friendly, lovely, obedient, smart and active animal. And in Chinese mythology, Hare is the spirit of the moon, living as a symbol of holiness, brightness, beauty and longevity and representing good luck and harmony. In westerners' eyes, the moon looks more like a cheese ball. And foreign people are telling their kids the Greek mythology of the Moon Goddess Artemis. And in China, people see a jade hare staying beside a rock at the foot of a cinnamon tree, holding the longevity medicine in its claws. As you know, a hare is a kind of animal that is reproductive and multiplies very quickly. A female hare can give birth up to 6 times a year and up to 6 newborns at a time. In China people believe the more kids,  the more blessings and they regard Hare as the Moon Goddess, showing respect to its incredible capability of reproduction. And in Chinese minds, the Hare is closely related to the lives and multiplying of human beings and people's beautiful hope about their futures.

Because of the natures mentioned above, the Hare has long been believed by Chinese people to be a smart and lucky animal. And the people who are born in the year of Hare are also considered as the luckiest among others. In each Spring Festival, of the year of Hare, people will make Hare shaped window paper decorations, hang new Hare year pictures, hang lanterns, Play lantern riddles related to the Hare and talk auspicious phases about the Hare to celebrate the coming year. And so will people point to the moon and tell their kids the stories of Chang E flying to the moon:

A Long time ago, there was an excellent archer, named Hou Yi. He obtained the herb of longevity from the West Queen, who is the wife of the Jade Emperor. The student of Hou Yi, Feng Meng, heard the news and tried to steal the herb. However, he was found by Chang E, wife of Hou Yi. Under the urgent circumstance, Chang E ate the herb and flew into the sky. As she was so in love with her husband and didn't want to leave, she finally decided to stay alone in the Moon Palace. In the tale, Chang E asked her servant, Wu Gang to cut the cinnamon tree and the Hare to grind the leaves, so as to make the medicine to help her return to her husband. But, unfortunately she and her husband still remained separated and missing each others company.

This is why, the Hare is considered an important animal in China.

② Legends and related literatures about Hare

In China, there are plenty of legends and literatures regarding the Hare. Here are some examples.

There are two versions of The Jade Hare Grinding the Medicine.

Legend 1: Three immortals turned into old and poor men and asked the Fox, the Monkey and the Hare for food. Both the Fox and the Monkey found food for them. However, when the Hare was asked, she said: "Please, you can eat me!" She then lept into the fire. The immortals were so moved by this they decided to send the Hare to the moon palace. From then on, the Hare has been staying with Chang E in the palace and helping her make the longevity medicine.

Legend 2: A Long time ago, two hares prayed to God for thousands of years and became immortals. They had four beautiful and lovely daughters. One day the male hare was called in by the Jade Emperor and had to sadly leave his wife and daughters, getting on the cloud for the heaven palace. By the South Heaven Gate, he saw Chang E passing by, escorted by the royal guards and the God of Venus. Feeling curious, he asked one of the Heaven Gate Keepers who they were, and was told the story of Chang E. He was very sympathetic towards the lady. However, his capabilities were limited. What could he do to help? The life in the Moon Palace must be so lonely and depressing. And Chang E should be surrounded by friends. The male hare suddenly remembered about his four daughters and he rushed back home. He told the story of Chang E to his wife and his thoughts about sending one of their daughters to the moon, the female hare also felt sorry for Chang E. However, she didn't want to sacrifice any of her children. Neither did any of the daughters want to leave. All of them cried sad tears. The male hare said earnestly: "if it were me being kept alone on the moon, would any of you come to stay with me? Chang E was trying to protect the longevity medicine and she is suffering so much. Don't you feel sad about her? Kids, we should not only think about ourselves." The daughters finally understood and were willing to help. With tears in their still crying eyes, the parent hares decided to send the youngest one. At last, the little hare left her parents and sisters and went to the moon to help Chang E grind the medicines.

Though in two different stories, the Hare is pictured as a warm-hearted and friendly animal in both. You can see the beautiful image about the Hare in Chinese people's minds. However, in the famous Chinese classic novel "Journey to the West", the writer pictured a different image about the Jade Hare. The Jade Hare in the Moon Palace was a mean and narrow-hearted animal. She turned herself into an Indian princess and tried to marry the Tang Monk. She finally failed in the fight against The Monkey King and was turned into Hare again. She was escorted back to the Moon Palace. This story tells people to always keep a nice and innocent heart, and to keep away from hatred and evil thoughts.

There is another beautiful story in Chinese folk history, named "The White Hare Story". Liu Zhiyuan was from a very poor family and his uncle was forced to join the army. His wife, Li Sanniang, was treated badly in the snobbish family of her mother in law. After their son was born, she asked somebody else to take the baby to Liu Zhiyuan and lost contact with them. Over ten years later, Liu Zhiyuan became rich through hard work. When his son went to the mountain to hunt one day, by following a white hare, he met with Li Sanniang. And the entire family got together again.

Besides these stories, there are some other idioms about the Hare, such as:

A wily hare has three burrows: it means that the wily hare usually has three burrows to hide. It figures that somebody has different places or methods to hide from their troubles.

Sitting by a stump, waiting for a careless hare: a farmer saw a running hare get caught in the net he put next to a stump in the ground, and get killed. He then put down the net and sat by the stump, in hope to get the second careless hare. It means that somebody wants to be successful without hard work and that somebody sticks to what he knows and never accepts change.

Crow flies and Hare runs: Crow refers to the sun and Hare refers to the moon. It means the passing of time.

Jade Hare and Silver Frog: it refers to the legendary Jade Hare and Silver Frog in the Moon Palace, picturing a beautiful and shining moon.

2. Introduction of Chinese New Year Custom

A.Mang Nian

Mang Nian, like a prelude to the Spring Festival, is the period between La Yue and the Spring Festival, when people are busy with preparations for the Spring Festival. In China there is the tradition of worshipping gods and ancestors at the end of the year when people used La Rou (salty preserved/smoked meat) as the sacrifice. Hence, La Yue, the twelfth month of the Chinese lunar year, was named after it with "La" month. In most parts of China, La Yue belongs to the winter time, serving as a bridge between the past and the future and the old and the new, since it is approaching to spring. During this period, people start to prepare for the Spring Festival. They are busy with house cleaning, shopping for food and clothes, contacting relatives and friends, arrangements for gatherings or booking trips to their hometown for the reunion gatherings.

a. Customs in the preceding days of the Spring Festival

① On the Eighth Day of La Yue, Eat La Ba Porridge

Strictly speaking, the first day of Mang Nian, falls on the Eighth Day of La Yue, known as "La Ba Festival" with the custom of eating La Ba Porridge.

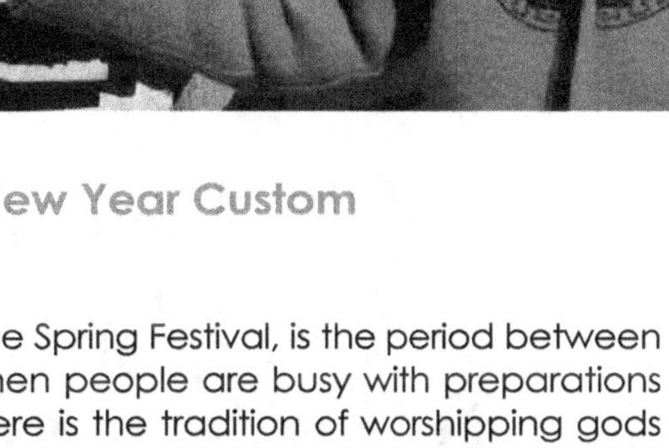

There is a saying in China "The chill on the seventh or eighth day of La Yue can even make one's chin drop". To resist the chill, the whole family therefore always enjoys a bowl of hot La Ba porridge around the table together. La Ba porridge usually contains eight ingredients such as yellow rice, glutinous rice, chestnuts, millet, red beans, Chinese dried dates and some others such as dried fruits or nut, though varying according to different areas in China. It need stewing for a long time until it becomes thick, soft and sticky and then can be mixed with sugar or salt. "Ba" (referring to eight) in La Ba porridge rhymes with "Fa" (prosperity), so to eat La Ba porridge which is made from eight ingredients on the eighth day of La Yue has the implication of auspiciousness and prosperity. In modern days, there are still some Chinese families who maintain the tradition of eating La Ba porridge on that day.

It is said that the tradition of eating La Ba porridge is related to Buddhism. According to the legend, Buddha Sakyamuni was the son of a king before he became a Buddha. He deeply felt the human being's afflictions from birth, age, illness and death etc, and then absorbed himself in to many years of an ascetic lifestyle. He travelled all the famous mountains and rivers throughout the world in a quest for enlightenment. On the eighth day of the 12th lunar month, exhausted from a long walk, he fainted on the roadside in a desolate place nearby the Nile. A shepherdess found him there and fed him porridge cooked with some food she brought with her, wild fruits and spring water. After eating the porridge, Sakyamuni came into consciousness. When he recovered, he jumped into the Nile for a purification bath and then he sat meditating under the bodhi tree where he realized the truth of Buddhism. With the introduction of Buddhism into China, the temples all over China have been holding ceremonies to commemorate him and offered porridge cooked with rice and fruit to Buddha like the shepherdess on that day. This is called La Ba porridge.

There is another legend about La Ba porridge. Once upon a time, there was a family of four with a couple and two of their sons. The couple was so hardworking that year after year that their barn was always stored with all sorts of abundant crops. Besides that they had a date tree in their yard, under the care of the couple, produced plenty of sweet and crispy dates. They sold out the fruits and earned quite a lot of money, which helped them to live a well-off life. As their two sons grew up, they were aging and in the end passed away. In their wills, the father told the sons to grow crops and the mother told the sons to take good care of the date tree. However, later on when the elder brother saw the full barn, he told his younger brother, "There is enough food in the barn for us, so we don't need to bother working this year." Alike his elder brother, when the younger brother saw the date tree filled with fruits, he said to his elder brother, " There are too many fruits to eat in the date tree. We don't need to do anything." In doing so, the brothers became more and more lazy and greedy, who enjoyed the life year by year and finished up their stock in the barn in a few years. The date tree became less and less productive due to lack of attendance.

On the eighth day of this La Yue, there was nothing left to eat, so the brothers had to search very carefully for food all over the barn. They fetched a broom and a dustpan and swept over every corner of the barn, where they got a handful of yellow rice here, another handful of red beans there, and even several dates. Eventually they gathered several handfuls of different kinds of crops, with which they cooked into porridge. While eating, they stared at each other and remembered their parents' words. They felt so regretful that they started immediately working from dawn to night like their parents had done so many years ago. Thereafter, they had a rich life again within a few years and then they lived happily with their families. To remember the lesson, people all followed the brothers to cook the porridge with different kinds of crops and named the porridge La Ba porridge.

As a kind of traditional Chinese festive food, La Ba porridge represents the dream of Chinese people for happiness. Nowadays, it is replaced by "Ba Bao (Eight Treasure) porridge" and became an ordinary food in daily life.

② Twenty-third day of La Yue, Worshiping the kitchen god

After the La Ba Festival, it starts teemed with a New Year atmosphere. Chinese people also call the Spring Festival as 'Da Nian (Big New Year)'

which is after "Xiao Nian (Little New Year)". The date of Little New Year varies depending on location. For example, it is celebrated on the twenty-third day of La Yue in the north of China, but on the twenty-fourth in the south. If La Ba Festival is regarded as the beginning of the prologue of Big New Year, Little New Year then is the climax of the prologue, after which Big New Year is counting down. Worshiping the kitchen god is one of the main activities during Little New Year, symbolizing the official start of the celebrations during of the Spring Festival.

Worshiping the kitchen god is also known as "seeing the kitchen god off". The kitchen god, named Zao Shen, Zao Jun or Zao Wangye, is a god who is in charge of the heart of every household. On the twenty-third day of La Yue he returns to the heavens to report the good and bad deeds of each family over the past year to the Jade Emperor who then gives either a reward or punishment to the family based on Zao Shen's report. Therefore, traditionally, every household would have a paper effigy of Zao Shen, sometimes with his wife named Zhao Wang Nai Nai together, above the fireplace in the kitchen. Couplets with the words such as "Report good deeds in heaven; Protect the family on earth" were posted beside the paper effigy, revealing their wishes so that Zao Shen could put in good words for them to Jade Emperor so as to bring good fortune to their families.

Thus, when seeing Zao Shen off, people would burn joss sticks first and then present offerings of sweet sticky food such as malt sugar and Chinese sticky cake, or even smear the lips of Zao Shen's paper effigy with honey to sweeten his words to Jade Emperor, or to keep his lips stuck together. After this, the effigy was burnt and firecrackers were lit to speed Zao Shen on his way to heaven and a week later replaced by a new one. Then one again on New Year's Eve followed by a ritual of burning joss sticks and bowing, which is known as "Welcoming the return of the kitchen god". Nowadays, most of the households in China do not worship the kitchen god any more, but the tradition of eating malt sugar is still popular in some rural areas.

There is a quite an amusing story about worshipping the kitchen god. It is said that in old times there were two brothers who got along with each other so well that they were unwilling to break up the family and live apart even when they got married. The elder brother named Zhang Dan was a bricklayer and the younger brother called Zhang Lin was a painter. Dan was adept at building cooking stoves which could save both labour and fuel, so his business thrived and he became well-known, honoured as "Kitchen god Zhang". Besides, he was very warmhearted. He always settled arguments between the neighbours through his convincing persuasions. Once there were disputes, the neighbours would always turn to him for help. Therefore, Kitchen god Zhang was admired by everyone. He lived 70 years and died at midnight on the twenty-third day of La Yue.

However, after his death, the family fell into chaos since he had been the head of the family and his younger brother, though in his 60s, had never run the family. The daughter-in-laws cried for dividing up family property so as to set up separate households, which bothered the painter a lot and made him feel gloomy afterwards. One day, a good idea finally occurred to him. On the one year anniversary of the death of Kitchen god Zhang, that is, the midnight of the 23rd day of La Yue, the painter woke up the whole family all of a sudden and told them that his elder brother showed his presence in the kitchen. Then he took them there where on the dark wall above the stove the dead Kitchen god Zhang and his wife emerged in the dim candle light. This shocked every one. The painter said, "I dreamed that my elder brother and sister-in-law had become immortal and been offered official posts by Jade Emperor. When he knew the discord among the family, he was very angry and planned to report to Jade Emperor and came to punish you on the New Year's Eve." The younger generations were so terrified at this that they knelt down to the Kitchen god Zhang and promised that they would never mention the breaking up of the family again. Moreover, they brought his favorite confection as sacrifice and begged for his forgiveness. Thereafter, the family became as harmonious as ever.

The news spread far and wide. When the neighbours heard about this, they came to the family to inquire about it. In fact, it was a portrait of Kitchen god Zhang that the painter drew and used to intimidate the younger family members. But he never expected that it would prove so effective. When the neightbours asked about it, he had to pretend that it was true and distributed the portraits to them. In this way, the portraits of Zhang Dan were posted in the kitchen of every household and worshipped as a real god. As years passed by, the custom of worshipping the Kitchen god was formed to wish for a happy and harmonious family.

As the saying goes, "Adults expect farming while children long for New Year". Little New Year is an exciting carnival for the children, because it signifies the happiest time of the Spring Festival is approaching. In ancient times, the common people lived a very poor life due to lack of resources. When the ceremony of worshipping the Kitchen god was over, children could enjoy the confection that they could rarely eat in daily life. Accordingly, children, who ran back and forth across their

houses from the kitchen to living room and outside, appeared the happiest and busiest during those times. However, the traditional atmosphere and fascination of the Spring Festival seems fading with the improvement of people's living standards. The children at present live a life with various affluent foods, clothes and other resources, as if they were enjoying the Spring Festival every day.

③ Twenty-fourth day of La Yue, Dust-sweeping

The custom of sweeping dust on the twenty-fourth day of La Yue has a long history. Chinese people attach great importance to the Spring Festival, so

all the households must thoroughly clean their dwellings to welcome the New Year day during the period between La Ba Festival and New Year's Eve. The Chinese character of "尘 (chen means dust)" is a homophone of "陈 (chen means antiquated)", as a result of which dust-sweeping before the New Year has an implied meaning of bidding farewell to the old and ushering in the new.

A legend about dust-sweeping says, there was a three-corpse god who was sent by Jade Emperor to find out how people obey the rules. Nevertheless, the

three-corpse god often framed up before the Jade Emperor that human beings deserved punishment since they often mobbed together and even cursed him. Therefore, The Jade Emperor got angry and ordered the three-corpse god to inspect every household on the twenty-fourth day of La Yue and mark the bad people's houses with their names covered by spider webs. Then, The Jade Emperor sent a heavenly general named Wang Ling Guan to carry out punishments with his scourge on the New Year's Eve. Once the marks were seen, the whole family would be killed. When the appalling news was heard by Zhao Shen, he decided to rescue the people. He told everyone to have a thorough clean through their houses from the twenty-fourth day to the thirtieth day of La Yue by removing spider webs and stains on the wall so that those name marks could not be found. In appreciation for Zhao Shen's help, people started annual house cleaning on the twenty-fourth day right after they saw

Zhao Shen off. Up to now, dust-sweeping is compulsory before the Spring Festival in China.

④ Twenty-fifth day of La Yue, Preparation for the New Year Goods

From the twenty-fifth day of La Yue Chinese people start to shop for New Year goods such as food, clothes, incense, candles and firecrackers for worshipping. It has almost become obligatory as well to prepare food before New Year's Day, such as fires and sharp instruments such as knives and scissors used for the preparations. Chinese people believe that a lack of preparation can provoke a mishap and thus ruin New Year 's Day.

Purchasing food plays a large role in the preparation process. In the past, butchering pigs was the most important deed because in most parts of China pork is served as the main course in the reunion dinner and the pig head as the major sacrifice to gods and ancestors. Apart from that, festive food includes Nian Gao or Ci Ba (Chinese sticky cake), fish, other kinds of meat and different types of vegetable, and confection

as well. When kids in the neighbourhood come over home for Bai Nian (New Year visits), they will be treated with confectionary such as preserved plums, candied dates, hawkthorn cake, peanut cake and walnut cake etc. With a variety of colourful confectionary, the houses are tinged with a happy festive atmosphere.

Shopping for clothes is a must-do before the New Year day as well. The stores are crowded with people who come to purchase new clothing for themselves and their families. Merchants take the chance to offer more new-year shopping discounts to attract customers so that they can make more profits and customers can buy cheap and satisfactory clothing. Festive joy flows out of everyone at this time of year and people will walk around with huge smiles on their faces.

⑤ Twenty-seventh day of La Yue, Washing away the dirt and bad luck

After shopping, Chinese people will purify and beautify themselves to welcome the New Year. They will bathe with argy wormwood juice that is believed to be able to completely cleanse the body and then get all their clothes and beddings washed and put on new fresh new clothes. The cleansing symbolizes the washing away of the dust, poverty and bad luck of the preceding year. Apart from that, there is another folk tradition of receiving a new-year haircut. As the saying "With or without money, receive a new-year haircut" goes, people will go for a new-year haircut

as well before the Spring Festival. Therefore, at the end of a year, the business of barber shops is booming. In this way goes the tradition of washing away the bad luck so as to present an entirely fresh high-spirited appearance to face the New Year.

⑥ Twenty-eighth day of La Yue, Decorating with paper-crafts, Spring Couplets, "the door gods" and New Year paintings

Decoration, another important festive activity, conveys a New Year greeting. Chinese paper-crafting is a unique folk artwork featured with skills of craftspeople, aesthetics and emotion, which can be used as a festive ornament and also conveys people's wishes for good fortune. After windows are cleaned, they are adorned with new paper-crafting 'window flowers' in different shapes like foursquare and circle etc., and in different themes such as portrayal of things in daily life, drama figures or twelve Zodiac animal signs etc. Although showing simple scenes, paper-crafting is usually charmingly bright as they are generally made using red paper, it is then exquisitely cut out of engravers' imagination.

Before the Spring Festival, couplets with poetic antithetical auspicious words written on red paper in the calligraphic style are posted on the front doors of every Chinese household, which is an important part of the celebration. The couplet is called a Chun Lian /Spring Couplet. Spring Couplet is

symmetrical and antithetical, normally with three words at the least in each line, such as vertical strips "天增岁月人增寿，春满乾坤福满门"(Another year is added to nature and humans; As spring comes into bloom in the universe and our familes) and the horizontal piece"万事如意"(All the best) or"五福临门" (Five fortunes come through the door). Spring Couplet has three lines, traditionally the first (upper) line posted on the right side of the door, the second (lower) line on the left side, and the third (horizontal) line on the top with the words written in the order from right to left. But in nowadays, the order can be inverted if the horizontal line is written from left to right. The red Spring Couplet can not only satisfy the superstitious beliefs of Chinese people but also symbolize life's renewal and prosperity in spring.

On the twenty-ninth or thirtieth day of La Yue, a kind of painting, known as "the door gods", is traditionally posted on the doorway to exorcise evil spirits and guard the household. The door gods are portrayed as two valiant heavenly generals named Shen Tu and Yu Lei, whose duty was to capture demons and ghosts harming people. It was said that on the East Sea in China there was a beautiful mountain, where a golden rooster crowed at the break of each day on the top of a 3,000 year old peach tree. In the north-east part of the tree was a branch bent down to the ground to form an arched door, through which thousands of demons and ghosts living in the mountain had to pass whenever coming into the world. The Jade Emperor worried about this and had two generals, named Shen Tu and Yu Lei, guard the door. Therefore, it was believed that the two gods were capable of subduing ghosts and that peach wood had the power to drive away evil spirits and ward off negative entities. As a result, the paintings with their images were posted on walls to keep evil spirits away and bring back peace and luck.

As time went by, the two gods were humanized into real historical characters such as the two generals in Tang Dynasty named Qin Shubao and Wei Chigong. Once, when Emperor Li Shiming (599 A.D—649 A.D) in Tang Dynasty was sick, he felt disturbed by the howling ghosts outside his door. Thereupon, two of his generals Qin Shubao and Wei Chigong guarded the door and scared off the ghosts. Emperor Li was pleased about it but unwilling to have them guard the door day and night, so he ordered to paste their portraits on the door and the ghosts never harassed him again. Soon the news was spread to everyone of the Tang Era, people all followed the suit and pasted the two generals' portraits on doors to guard their households.

In addition, Zhong Kui is another Chinese famous folk door god. It starts with the Emperor Li Longji (685 A.D–762 A.D) of the Tang Dynasty, being sick in bed. One night, he dreamed about a devilkin stealing his jade flute. Just then, a tall ghost with a shabby hat and a green robe appeared and caught the devilkin, and then tore out his eyes and ate them. The Emperor asked who the tall ghost was. He said his name was Zhong Kui. When the Emperor awoke, he had miraculously recovered from his illness. He ordered his court painter to draw out the image of Zhong Kui in his dream: an ugly appearance with messy hair, a green robe and a black hat and black boots, raising his eyebrows and staring in anger, catching a devilkin with a sword in his hand. This is the famous 'Painting of Zhong Kui Catching a Ghost'. Ordinary people followed the Emperor to respect Zhong Kui as a door god and place his portrait on doors, hoping to drive away the evil spirits and gain safety.

The door is the division between the household and the outer world. When people feel that evil or negative entities cannot be driven away by themselves, psychological dependence is needed to maintain the sense of security. Chinese people rely on the door gods to keep away the evil and ghosts and ward off disasters, so on the twenty-ninth or thirtieth day of La Yue, old paintings of door gods are replaced by new ones to regain the protection of door gods.

Chinese people like to hang up New Year pictures in their living rooms and bedrooms as well. A New Years picture (Nian Hua in Chinese), is a popular folk artwork in China, reflecting folk customs and beliefs and hopes for future. It originated from a painting of the door gods but was added with more subjects later on, such as fairy tales, historic stories, landscapes and flowers etc. For example, large numbers of famous colourful New Year pictures were produced like 'Picture of three gods of Good Fortune, High Salary and Longevity', 'Good fortune offered by heavenly officials' 'Prosperity in Crops' etc. Although the traditional colour of a New Year picture to drive away the negative entities has faded away, it is still popular in households, used as ornaments to express the desires for safety and wealth.

In the Chinese lunar calendar, December 30th of each year is called the "Eve"(The Chinese Lunar December only has 30 days or 29 days, and we uniformly name the last day the "Eve"). The Reunion Day is on the eve of Chinese New Year. All the people who are traveling or living abroad are expected to return home to reunite with their family before that day, unless they have really important issues to attend to. Therefore, the Reunion means getting all the traveling family members back home to spend the festival together. Lots of activities take place on that day. People prepare different kinds of tasty dishes and drinks and worship the heavens, the gods and their ancestors. People will also remember the God of the Kitchen. After that, every family will start the reunion dinner, enjoy the scrumptious food and welcome in the New Year.

a. New Year Custom

① Post the character of "Luck"

In China, the character of "Luck" means happiness and fortune. In the Spring Festival, people would post it on their doors, to show their desire for happy and peaceful lives. More interestingly, people often post the character up side down, to show that luck, fortune or happiness has been bestowed by the Gods down to earth from the heavens.

There is an old folk tale about the character for Luck. During the Ming Dynasty, there was an emperor, named Zhu Yuanzhang (1328AD-1398AD). One year, on January 1st, he disguised himself as an ordinary person and traveled to the small town of Huaixi (West of Anhui Province, China). There he found a big group of people looking at a cartoon drawing of a barefoot lady holding a watermelon. The cartoon was making fun of women with big feet (In ancient China people thought it was

ugly to have big feet if you were a women.). However, Zhu Yuanzhang's wife was from Huaixi and had big feet herself. He felt annoyed and thought that these people had damaged the image of the imperial family. Then he sent the local officials to investigate and post the character of 'Luck' on the doors of all the people who had not seen the cartoon. He then sent a small army to capture the families without the Luck character posted on their door. But the Queen was a gentle and kind lady. After finding out what her husband had done, she decided to help the people avoid such a cruel punishment. That night, she sent some workers to the little town and told every family to post the character for Luck on their door. On the second morning, when the army arrived at the town, they surprisingly found that every family had posted the character on their doors. Zhu Yuanzhang was confused and finally learned that it was the Queen helping the people in the town. Though feeling angry, he started to understand what the Queen meant. However, one family posted the character up side down, as no one in this family had ever seen this character before. With great anger, the emperor ordered everyone in that family to be killed. Thinking quickly, the Queen smiled and told him: My Majesty, don't be angry. I think the people in that family regard you as the God of Luck. They knew that you would arrive today and posted the character upside down intentionally. Look at it! It means that the Luck is bestowed by the Gods to the earth." (Notes: in Chinese the character of 'upside down' has the same pronunciation as Arrival) Hearing this, the emperor felt very delighted and ordered for the family to be instantly released. A disaster was avoided by some quick thinking from the Queen. From then on, on every New Year Eve, Chinese people post the character up side down to remember how the Queen had saved lives. As time passed, people used this tradition to pray for good luck, rather than to avoid trouble. People still continue this tradition today.

② Worshipping ancestors

In China, worshipping ancestors is one of the most important activities in the New Year. Chinese still pay great respect to the dead. Before New

Years Eve, every family will put the tablets or statues of ancestors in the main halls of their houses. Fruits, cakes, candles, drinks and so on will be left around the statues. At about 11 pm on the Eve, all the family members will dress up, bow down and pay respect with incense, in accordance with age order. Everything is kept formal and quite serious out of respect. Worshipping ancestors is not just about showing respect, but also praying for the luck and fortune for the coming year. Now most Chinese celebrate by having simpler activities to worship ancestors with food and flowers at their tombs, to carry on with the traditions and virtues.

④ Staying overnight and paying 'Lucky Money'

After the New Year Eve dinner, every family will close the door and sit together for the whole night without sleeping. All the lights should be kept on until the morning. It is formally called staying overnight. In such activities of the New Year, people will put snacks and fruits on the table, as well as some lucky fruit, such as apples (the

name of apple in Chinese sounds like peace). And some families will cook a big bowl of staple food with rice and wheat to represent a good fortune with a lot of gold and silver. It also means that the family will have enough, or even more than enough food, for the next year. After that, the whole family will sit together, eat food and have fun, waiting for the next day and the New Year.

'Lucky Money' is one of the necessary gifts for the New Year. Every Chinese child deems it as the most joyous thing in the Spring Festival. When visiting the older generations, the younger generations will receive the prepared Lucky Money from their aged relatives. It is an important New Year custom in China, carrying the best wishes from the older generations. This money is given as a token of good luck to the kids. Chinese people believe that kids are easily attacked by evils and diseases. So the money is considered as a blessing for health and luck to children. Generally the older generations will pack the money in brand new red envelopes and give it to the kids during the New Year Eve visits. Some parents also hide the envelopes under the pillows or the sheet of the bed while the children are sleeping, mimicking the way Santa delivers gifts in the night to western children. It is a big surprise every year for the children.

There is a legend about Staying Overnight and the origins of 'Lucky Money'. In the legend, there was a crude monster, named "Sui", with a black body and white hands. Every New Years Eve, the monster will sneak into each families home and touch the heads of the sleeping kids, to scare them and to infect them with disease. People were extremely fearful of the monster. Therefore, in order to protect the kids from such torture, every Eve after dinner, the family members will sit together and keep awake till the next morning. Therefore, Chinese Staying Overnight is named "Protecting from Sui". However, Sui still came to hurt children every New Year Eve. Once there was an old couple, who got their first baby very late on in their lives. As a result, they cared very tenderly towards their child. During the New Years Eve, both of them worried so much about their baby being hurt by Sui so much, neither of them could eat during dinner. After that, they began to stay awake overnight with the child, with the lights turned on. The child played different kinds of toys and finally asked for a piece of red paper and 8 coins. He packed and unpacked the coins with the paper repeatedly before finally exhausting himself. Yawning, he then put the red pack with coins besides his pillow and fell asleep. In the middle of the night, the old couple felt increasingly tired and finally fell into a deep sleep. A few hours after midnight, it suddenly started to become increasingly windy outside and blew the front door of the house wide open and the gust of wind quickly blew the light out. In the house, slowly crept the monster, Sui. It stretched out its hands and pulled aside the bed curtain. It reached out a hand to touch the head of the child. However, a sudden light flashed from the side of the pillow and scared its hands away. Screaming, Sui immediately fled from the house. When the couple woke up, realizing they had let themselves sleep, they quickly lit up the room and found their child safe and still in a sweet sleep, like nothing had happened. They were so puzzled about what could have scared the monster away. They searched the bed and only found the red pack of coins. Could it be the red pack that scared Sui away? On the next day the old couple told their story to everybody they met and the story spread quickly. People then started to believe that the red envelope could keep Sui away from their children, and from then on, everybody followed the same method. Packing 8 coins into red paper and giving it to the children after the New Year Dinner. The pack was then placed beside the pillow. Surprisingly, year after year, Sui never appeared again. This custom was then carried on and developed into the modern day tradition. Since the coins packed in red paper could protect children from Sui, people started to call it the Lucky Money, or in Chinese the "Money to keep Sui away". As Sui has the same pronunciation as "Year" in Chinese, people changed the name to "Money to protect the year", meaning 'Lucky Money'.

⑤ Firecrackers and hanging lanterns

Chinese people believe that the sound of firecrackers can not only expel the evils but also protect the peace of the whole world. It also increases the joy of the New Year. Also the smoke of firecrackers can kill the bacteria in the air. In the New Year days, Chinese people will enjoy the beautiful scenery of firecrackers in the sky, as well as the great feeling of excitement. As the New Year gets closer, there will be more and more people setting off firecrackers. The idea is that on New Year Eve, the

dgxnwhole earth will be shaken by the sound of firecrackers and covered in smoke. Then when the host of the Spring Festival Show announces the beginning of the new year at 12am, the entire country will begin to celebrate. It is also a national carnival for all Chinese. The fireworks and red cracker ashes will bless every Chinese to pass the year with happiness and luck.

As a traditional Chinese art craft, red lanterns represent family unification, luck, happiness, brightness and hope. Therefore, in each Spring Festival, every family in China will hang red lanterns in front of their houses. It not only creates a colorful picture on the eve of the new year, but also bears the great and beautiful hopes of people to their futures. Especially at night, every city and village of China will be lit up by the red lanterns, and the red light of the lantern makes every body, locals or visitors feel like staying at home under the warm glow.

b. New Year Foods

① New Year Dinner

asked for a piece of red paper and 8 coins. He packed and unpacked the coins with the paper repeatedly before finally exhausting himself. Yawning, he then put the red pack with coins besides his pillow and fell asleep. In the middle of the night, the old couple felt increasingly tired and finally fell into a deep sleep. A few hours after midnight, it suddenly started to become increasingly windy outside and blew the front door of the house wide open and the gust of wind quickly blew the light out. In the house, slowly crept the monster, Sui. It stretched out its hands and pulled aside the bed curtain. It reached out a hand to touch the head of the child. However, a sudden light flashed from the side of the pillow and scared its hands away. Screaming, Sui immediately fled from the house. When the couple woke up, realizing they had let themselves sleep, they quickly lit up the room and found their child safe and still in a sweet sleep, like nothing had happened. They were so puzzled about what could have scared the monster away. They searched the bed and only found the red pack of coins. Could it be the red pack that scared Sui away? On the next day the old couple told their story to everybody they met and the story spread quickly. People then started to believe that the red envelope could keep Sui away from their children, and from then on, everybody followed the same method. Packing 8 coins into red paper and giving it to the children after the New Year Dinner. The pack was then placed beside the pillow. Surprisingly, year after year, Sui never appeared again. This custom was then carried on and developed into the modern day tradition. Since the coins packed in red paper could protect children from Sui, people started to call it the Lucky Money, or in Chinese the "Money to keep Sui away". As Sui has the same pronunciation as "Year" in Chinese, people changed the name to "Money to protect the year", meaning 'Lucky Money'.

⑤ Firecrackers and hanging lanterns

Chinese people believe that the sound of firecrackers can not only expel the evils but also protect the peace of the whole world. It also increases the joy of the New Year. Also the smoke of firecrackers can kill the bacteria in the air. In the New Year days, Chinese people will enjoy the beautiful scenery of firecrackers in the sky, as well as the great feeling of excitement. As the New Year gets closer, there will be more and more people setting off firecrackers. The idea is that on New Year Eve, the

complaining about their ears being seriously hurt. He felt terrible to see people suffer like this. Then he asked his assistants to build a big shelter in the neighborhood. They put a huge pot in the middle of that place to make medicines for the poor people. One of the medicines was called the "Cold Resisting Dumpling Soup", made of lamb, peppers and some other cold resisting herbs. Once the ingredients were cooked, they minced and wrapped them with flour dough, in the shape of ears. The dumplings were boiled and donated to the poor people. By eating these, people felt warm and relaxed. The pain on their ears was also getting better and better and finally recovered. From then on, every January 1st, to celebrate the New Year as well as to memorize the cure for the people who felt such pain, people started to make dumplings, which are usually served on the morning of that day. And people named this food "Dumpling" to memorize the famous and warm hearted doctor, Zhang Zhongjing.

Zhang Zhongjing lived 1800 years ago, and the story about him and dumplings is widely known among Chinese people. On the other hand, since the shape of dumpling looks like a golden ingot, it also gives people expectations for good fortunes in the next year. As time passed, dumplings have become a popular New Year treat.

As people know, dumplings represent an important part of Chinese culture. There are different kinds of fillings, like pork, lamb, beef, fish, triple flavors and so on. And the fillings of the New Year Dumpling are just as interesting. People put

candy, dates, chestnuts, peanuts etc. in them. Some families even put coins in dumplings. Candy represents a sweet life in the next year. Dates repreesent a hard working year. Chestnuts represent an exciting and busy year. Peanuts represents longevity and good health. Coins represent a whole year with good fortune. No matter what kind of special fillings people find, the other members will give him sincere wishes for the next year. If the children find the coins in the dumplings, the grownups believe this a sign of good luck for their child. They will then give them red envelopes to their child as a form of encouragement.

③ New Year Breads in different shapes

During the Spring Festival, most country families in north China will make steamed New Year Breads. It is a traditional steamed food in north China, representing a better life in the coming year. Now, the bread is used not only for worshiping gods and ancestors, but also as an important gift and food in all kinds of spring festival visits. Made from fermented flour, the bread will be made into different shapes, such as flowers, birds, fish,

fruits, zodiac animals and other interesting shapes. Due to the special texture of flour and the ornamentation of beans, the look of the bread becomes as important as the taste. People will paint the bread with food colouring to make it an artistic delicacy.

④ Sticky Rice Cake

In many places in China, people eat sticky rice cakes in the Spring Festival. It is steamed food made from sticky rice. After cooking, it will be pounded and cut into bars of different size. After drying, it can be served as a New Year Food. People make different flavors of cakes, but most families make it sweet, steamed or fried, in the south and north of China. In south China, besides steaming and frying, people also cook it in other ways, such as boiling and stir-frying. It actually can be made sweet or salty. In the Spring Festival, people also eat it together in a type of hot pot. It represents a good fortune and better lifestyle in the next year. Chinese people also make it into the shape of coins, Ruyi and so on, as a symbol of luck and fortune. And some families make the cake into the Jade Hare or other animals, representing longevity and luck.

⑤ Ciba

In south China, Ciba is a traditional home snack and treat for visitors in the Spring Festival. At the end of each year, every family will be busy making Ciba. People steam the wet sticky rice and put the hot dough into a special container. Then it will be pounded on, after this is done, people will take it out and put it on a big chopping board, which is covered by rice powder. When it cools down, it will be cut into different shapes, round or square, and preserved for future use. In the Spring Festival, Ciba can be preserved for more than 2 weeks in water, and the water will be changed regularly to keep it fresh.

C. New Year Visits

When the rising sun sheds first light onto the earth, everything is believed to look fresh and new, and the New Year washes over the people. In the Chinese lunar calendar, the first month is called January and the first day of January is named January 1st, or The First Day, The First Morning. New Year Visits are the most important priority on January 1st. The visit is also known as the greeting of the year. Though it doesn't have to be on January 1st, it starts from the first day of the lunar calendar.

a. New Year Traditional Customs of China

① January 1st, Opening the door, playing with firecrackers, worshipping ancestors and New Year Visits.

If New Years Eve is a farewell day to the passing year, January 1st is the beginning of the New Year. On that day, at the crowing of a rooster, Chinese people will get up to set off firecrackers to celebrate the beginning of the next year. The clear and loud sound of firecrackers opens a brand new year and takes away all the bad luck and sadness of the last year with the smoke. Therefore, in all of China, the first thing to do on the first day of the year is set off firecrackers, also named, opening crackers. In many parts of China, people believe that the first family to fire the crackers will get the first bit of luck of that year, and the first day of the New Year will be passed in the exciting sound of fireworks for every family.

Worshiping ancestors is the core content of the New Year activities. Chinese people believe that during the spring festival gods from different areas will also come back to the human world. Therefore, people will present gift snacks like fruits, breads, sticky rice cakes and so on at their homes. They will also dress in new clothes, accept their gods, worship the gods, and pray for the coming year and worship ancestors. After that, people will start the New Year Visits to each other.

New Year Visit is an important customary activity in China that has always been a tradition. You may still remember the legend of the "Year". After a whole night staying at home, with the door closed, red paper posted and firecrackers set off, on the first morning of January 1st, people will go out and greet each other. They congratulate each other for passing the night safely and staying away from the monster. As the years have passed, this activity has become a traditional custom in China.

People will finish the dumplings of New Year Eve for breakfast on January 1st and visit each other to give their best wishes. There are two kinds of visits in Chinese tradition. One is to visit the older generations. The young people will kowtow to the aged and wish them longevity. The older people will give Lucky Money to the younger generations, for health and good luck. Then people will chat to give each other encouragement and good wishes for the coming year. The other kind is the visit to friends and neighbors. People wish each other good luck and happiness for the whole year and so on. When visited, the hosts will treat the visitors with snacks such as wine, candy, sweet cakes, honey and dates etc. It means the New Year will be sweet (candy), improved (cake), with new babies (date).

To give wishes to friends or relatives living far away, or in case that the family has so many friends that they don't have time to visit all of them, people will send out New Year cards. In ancient China, people wrote their names, address, official titles and the words of best wishes on pieces of wood or paper and had someone to deliver it. It was an easy but very good way to send out best wishes. New Year Visit is not only regarded as a fun activity to share the joys of the Spring Festival, but also an important way to strengthen the connection between different families and different members.

② January 2nd, worshiping the God of Fortune, Returning to Your Parents House (Wife's parents)

Money is the material base of a happy life. Therefore, the god of fortune is regarded as a very important god in China. In most parts of China, on January 2nd, people will worship the god of fortune. In the early morning, people will light up candles and present wines, live roosters and live carp in front of the gods' pictures. After that, people will set off firecrackers and paste the picture on the wall of their sitting rooms. That means the god has been invited into the family. The street vendors will take this business opportunity to make the god statues and pictures and sell it to the surrounding neighborhoods. People should not say "No, I don't need that" if they don't want to buy it, but "We already have that." Otherwise, the god and the good fortune will be driven away. The worshiping of the god of fortune is to try and bring on luck and good fortune for the next year.

In most places in China, wives will visit their parents on January 2nd with their husbands and kids. They will also take presents with them. It is called Returning to the Parents House, and there are different requirements about the presents. The gifts should be in even number, representing good luck, and the parents will also prepare good food when visited by their daughters and sons in-law. They will also pray and give best wishes for the kids in the New Year.

③ January 5th, Opening of Business

In China, the business opening day during the New Year period is very important day for businesses. Usually stores will be opened on January 5th and the opening time is usually set at 8:08am, 8:18am, 8:28am and so on which are considered lucky numbers to Chinese. People will set off firecrackers, hang colorful lights and play music in hope that the business will be prosperous with excellent profits in the next year. The first customer will usually be treated warmly, with different kinds of nice snacks and an attractive discount. But now the public holiday of Spring Festival has been prolonged until 7th or 8th. As a result, the opening times of businesses are also changed accordingly. However, the custom of setting off firecrackers is well kept.

b. New Year Entertainment

① Dragon light dance

The Dragon is regarded in China as a holy animal, which has the supernatural abilities to control the whether, to diminish diseases and disasters and to bring luck and peaces to the human world. It is a very important part of Chinese traditional culture. China is a country carrying the culture of the Dragon, and people celebrate and pray for peace and a good harvest by performing the dragon dance. The dragon dance has become a necessary activity for the Spring Festival.

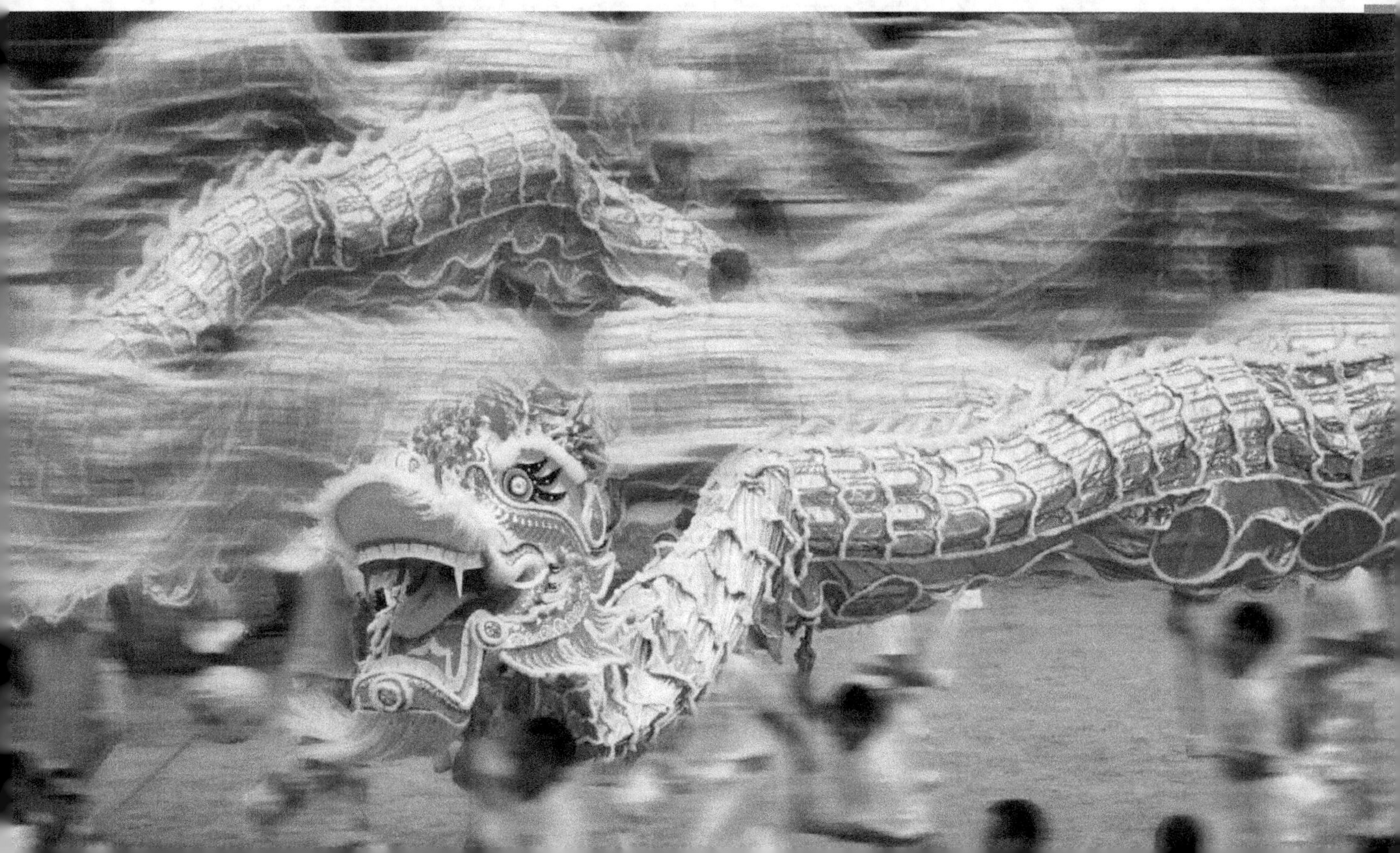

Usually, the dragon dance is made by a team of 11 or 13 players. The longest team could be composed of over 100 people. In front of the team, there will be a person handing a colorful cloth ball and leading the team. The dragon will follow and try to catch the ball, winding up and down and shaking it head and tail. On different occasions, the scale of such game will be different. Drums and gongs will be played according to the movements of the dragon. The Dragon Light Dance is so named because the dragon dance is usually combined with beautiful lights, and in some places of China, the dragons are actually made of lights. While the dance is being played, there will be other colorful lights around. If the activity is held at night, people will put fireworks into bamboo or iron pipes and the fire sparks onto the body of the dragon from different angles. The dancers have to maneuver the dragon quickly to keep the sparks off the dragon. The less sparks to touch the body of the dragon, the higher the skill the players have. Therefore, during the game, every player is excited and they are all regarded by people as the heroes.

② Lion Dance

In Chinese minds, a lion is a symbol of bravery and luck. And it has the ability to expel the evils and protect people. As a result, people used to perform lion dances in Spring Festival or other important activities, to pray for good luck and a peaceful life. There is an old story about how the lion dance came about. During the time of Song Wen Emperor of the South Dynasty (407AD-453AD), an official was ordered to fight against the empires enemies. But unfortunately he lost the battle, because the enemy soldiers were riding elephants with long weapons in hands, while his troops were mainly army men with light weapons. His troops were completely disadvantaged in the battles. One day, one of his generals advised him that the lion was the king of all animals and elephants would be fearful of them. Then the official asked his soldiers to make many "lions" with leather and cloth overnight. The "lions" were painted into different colors and theirs mouths were open. Every "lion" was driven by two soldiers. They hid in the grass around the battlefield, and the soldiers also built many traps around themselves. When the enemy elephants started to attack, the soldiers sent out the "lions". The sharp teeth and paws of the lions scared the elephants, and the elephants began to flee in different directions. Many elephants fell into the traps, and the South Dynasty finally won the war. From then on, the Lion Dance has become a folk custom of Chinese people and an important celebration for different festivals and important days throughout the Chinese calendar.

In the dance, there is a big lion and a little lion. The little lion is played by one person and the big one is played by two. One person will control the lion head while the other person will control the lion body and tail. The lion player wears the lion suit, and trousers of the same color as the lion and the golden color paw boots. The appearances of the players are coherent with the lion and the viewers just can't distinguish the players from the lion. Other than the lion, there will be a lion guide, who is dressed as an ancient warrior, with a Buddha mask, handing a fan or cloth ball, to guide the lion and the dance will be played in the sounds of Chinese folk music, by drums, gongs and so on. As guided, the lion will play different difficult acts like tumbling, jumping, channeling the table, standing on balls, walking on plum piles and so on, and making the dance interesting and lively.

③ Temple Fair

Temple Fair, originated from the worshipping activities of ancient China. It was mainly set as an entertainment activity to attract more believers and audience for different religious tribes. Now the Temple Fair of China has partly become a trade fair, with different activities of culture, trade and entertainment. It is no longer about religion. Going to Temple Fair has become an important and customary thing to do for Spring Festival in many places in China. It starts on January 1st. The temple will be full of visitors and vendors. Different types of goods can be found here, like garments, food, antiques, calligraphy, paintings, flowers and pets and so on. Acrobatics, shadow puppets, blowing sugar figures, clay figurines, dance, drama, entertainment and all kinds of local food from different places of China can be found here. A great many people will join this activity. It is a fun thing to do in the Spring Festival and an important social activity for people to communicate and do business.

④ Watching Operas Shows.

In previous years, people didn't have much entertainment. One of the a few interesting things to do is to watch the opera shows in the Spring Festival, after they finished the agricultural works. The opera troupe usually would set up a stage in the middle of the business area and present their excellent performances from January 1st to January 15th. On the stage, is the music of drum and gong, the actors and actresses sang beautiful melodies and under the stage people and families from the neighborhoods watched with great excitement. In the New Year, people would free themselves from the hard agricultural works, get together with friends and greet each other. It has become part of the New Year Visit and Reunion. Young people would take this chance to get to know each other and to show their love to their partner. Therefore, watching opera show has become one of the biggest events for Chinese in the Spring Festival, and now the out door stage has been replaced by the wide and beautiful theater. The folk troupe has also changed to professional opera show companies. Most people will sit in front of TV and watch the Spring Festival Evening, as well as other shows, with great anticipation.

⑤ Yangko Dance

The Yangko Dance was invented in the activities of transplantation and now it has become one of the important forms of entertainment during the Spring Festival from January 1st to January 15th for some parts of China, especially north China. The Yangko team is usually composed of 10 to over 100 members. They dress up as historical figures from stories, myths or real life. During the drums and music, the dancers will dance with beautiful poses and different team shapes. It is a really interesting activity and loved by many Chinese people. Now, in each Spring Festival, different Yangko teams will show their performance in the streets, greeting each other, passing on their good wishes and increase the happiness of the New Year.

⑥ Stilts-walking

Stilts-walking is a show performed by ordinary people during the Spring Festival, Lantern Festival and so on. The stilts are made of two plain wooden poles, bound on the legs of the performer, and the performer will stand on them and present the shows while walking. The wooden stilts can be as tall as 250cm. The mid-sized ones are over 120cm while the short one should be at least over 30cm. Different figures in the old Chinese myths will be performed as, in different dresses. In folk music, opera or the sounds of drums and gongs, the dancers will walk ahead and perform. It's a really funny and attractive event for visitors.

There is a story about the origins of Stilts-walking. In the past, people would get together and hold different kinds of New Year activities during each Spring Festival. However, one year, a local administrator came to govern the whole county, and he was a corruptive man. He did everything he could to take the money out of the pockets of people. He announced that the New Year activity would bring evil spirits into the town and the West, South and North Gates of the town would be closed. So the activities should only be held through a suspension bridge on the moat of the East Gate and people had to pay tolls to pass the bridge. This policy made all the people extremely upset. But soon they figured out a clever plan to deal with it. They bound two long poles on their legs and walked across the moat. By knowing this, the local administrator went out to check thee festival and he was pushed off the town walls and fell into the moat, by the crowd ot curious people, who had crowded here to see what had happened. His assistant lowered the suspension bridge to save the local administrator. At this moment, people waiting outside the town rushed forward into the town. From then on, Stilts-walking became a regular activity during the Spring Festival.

⑦ Mahjong

During the Spring Festival, Chinese people like to get together with friends and relatives to play a Mahjong. The rules of Mahjong are very simple and easy to learn, but with many varieties of playing styles. Different numbers of people can play different types of Mahjong. Therefore this game has been widely appreciated by Chinese people. In the New Year, the entire family will sit together to play Mahjong and chat, and always have a lot of fun. It is not only an entertaining activity, but also an important way to strengthen the connections among family members. It is now regarded as an Indispensable form of entertainment among Chinese people during the Spring Festival.

D. Lantern Festival

Lantern Festival is an important traditional festival of China, also named Shang Yuan, Yuan Xi or Yue Ye, because it is regarded as the first full-moon night of the New Year. In Chinese custom, people take the lantern festival as the last day of the Spring Festival. While the joy of New Year continues, people feel differently on that day, as the Spring Festival is ending at this day. Therefore people will take this last day to travel, to watch the lantern shows, to play quiz games and to eat sticky rice balls. The single people will also take this opportunity to get to know other people and to hopefully start an unforgettable love. As a result, the Lantern Festival is also called the Chinese Valentine's Day.

a. Legends of the Lantern Festival

There is a story about the origins of the Lantern Festival. The Hanwu emperor of Han Dynasty has a knowledgeable and warm-hearted minister, named Dongfang Shuo. One day, Dongfang Shuo saw an odalisque jumping into the well. He stopped her and learned that her name was Yuan Xiao. She had been working in the palace for a long time and did not have a chance to take care her parents. She recently learned that her parents were suffering from a serious illness. It made her miss her home badly. Feeling so depressed, she wanted to end her life by jumping into the well. Dongfang Shuo felt so sympathetic towards her and agreed to help her in seeing her parents. Then Dongfang Shuo had his servants spread a rumor that the Jade Emperor was angry, and had ordered the God of Fire to burn the entire city on January 15th. This news made people living in town very nervous. The Hanwu Emperor was also scared after hearing this. He asked Dongfang Shuo for advice. Dongfang Shuo said: "the God of Fire likes to eat sticky rice balls, and the odalisque Yuan Xiao is very good at cooking such food. You can send her out to teach people to make sticky rice balls by January 15th. Then the God of Fire will be delighted when presented with this food, and the town will be saved. At the same time, we should have the people to hang lanterns and to set off fireworks. The God of Fire then won't cause us any trouble." According to the order, the odalisque was sent out of the palace and she went back to see her parents. People of the town started to hang lanterns and set off fireworks. All the streets were lit up. People enjoyed the lantern activities so much and the festival passed with a great success. From then on, on January 15th, people will make sticky rice balls, set off firecrackers and hang lanterns. As Yuan Xiao was so good at making sticky rice balls, the festival was named as the Yuan Xiao Festival (Lantern Festival).

b. Special Foods of the Lantern Festival

① Watching Lanterns

The lantern show is a typical festive activity during the Lantern Festival. In ancient China, the lanterns were made of azure stones of different colors and different sizes. Pictures of flowers, birds, water, mountains and people were drawn on them. On the night of the festival, all the streets will be hung with different kinds of lanterns, which illuminate all the streets. After that, special markets selling lanterns were developed, and people started to buy lanterns, hanging them for all to see. As time passed, it has become a set event of the festival.

What's more, another traditional and famous custom of the Lantern Festival is to fly the Kong Ming lantern. The lantern was said to be invented by Zhuge Kongming (181AD – 234AD) in the Three Kingdoms Period. In a battle against Sima Yi, Zhuge Kongming was trapped and surrounded by the enemy. No one could get out of the town for reinforcements. Kongming examined the wind and made lanterns, on which contained messages to the government for reinforcements. The lanterns were carried by the wind. It saved the entire army, and the lantern has been named after Kongming from that day onwards.

The invention of Kongming Lanterns shows the intelligence of thee ancient Chinese. The frame of the lantern is formed by bamboo, the same as a kite. The lantern body is covered by paper, painted with wood oil, to retain the hot air. An oil lamp is put in the center of the frame. Once lighted, the fire raises the air temperature, higher than the air outside the lantern. This decreases the gravitational force of the lantern. When it's set off, it will fly into the sky on an early spring night, like a shining star.

② Playing Quiz Games

Quiz Games are smart game invented by ordinary Chinese people. It is also regarded as a traditional art of China. People paste quizzes on lanterns of different colors for fun. The person who answers the quiz correctly will obtain different kinds of little gifts. So it is also called the Lantern Quiz. The quiz is usually composed of simple languages, and the answer should be matched in every word. The nature, quality, function and so on should be hidden and implied in the quiz. It is a smart, literary and funny game to play. The winner should also be smart, logical and have a good imagination. It is a favorite game for many people during the period of the Lantern Festival. For

example: "It's like gourd and pear, it turns everything bright everywhere; without it the night is dark with fear, and can't see a thing here nor there" (Light Bulb); or "mountains with no trees; rivers with no water; streets that can't be walked on; it is the smallest but also the largest world" (Map).

③ Eating Sticky Rice Balls

Eating sticky rice balls has become a must for January 15th since thee earliest days of ancient China. The sticky rice balls are made of sticky rice powder. It can be as big as a walnut or as small as a pearl. Depending on whether there is a filling, it can be divided into two types. The flavor could be sweet, salty, spicy or sour and the fillings could be made of sugar or vegetables. The sweet fillings include walnuts, sesame, special types of plant, date paste, red beans, peanuts, etc, and they can be boiled, fried or steamed. But the most common way of cooking is to boil them.

3. Changes to Chinese Spring Festival Customs

With the rapid growth of the Chinese economy, the development of new cultures and the general improvement of people's living standards, the customs of the Spring Festival are being changed. Some old customs have been replaced by the modern forms of entertainment. Some others have been simplified and changed according to the interests of modern lifestyles, and some new customs, which combine both the traditions and modern ways of entertainments, are created. This reflects the cultural change of modern Chinese society.

The changes to New Year Visits. In the past, the younger generations would visit their elders during the New Year. They would even bow to their elders. As the continuous development of telephones and the Internet, the methods of New Year Visits have changed drastically. People are making greeting calls, sending text messages and emails. Now, the New Year text messages have become a necessary part of the New Years Activities, along with firecrackers and New Years Dinner, for modern day China.

The changes to New Year Gifts. Visiting relatives and friends during the Spring Festival is a traditional activity. Chinese people used to take candy, fruit, cigarettes, wine and so on as gifts, to send their best wishes to relatives and friends. Now, fresh vegetables have been used by more and more people as the New Year Gifts and they are widely appreciated.

People put the fresh vegetables in the boxes, which is then tied with a bow, to make a beautiful gift for their friends. Gifts are just a token of friendship and best wishes. Though vegetables are not expensive, such gifts are much healthier than cigarettes or wine.

The changes to New Years Dinner. No matter how busy people are, they will cancel all kinds of working appointments and spend time with the families during the Spring Festival. All family members will sit together to have New Years Dinner and to enjoy the comfort of staying with relatives and the fun times they have. In the past, people usually had dinner at home. But as time changes, more and more Chinese families, especially the ones living in big cities, choose to eat at good restaurants, as the environment and food are both much more pleasant, as well as the service. People don't need to start preparing for the dinner several days beforehand, and different families can even sit at the same table to enjoy their dinner. Adults can get together and chat about memories and kids are also happy to play together with their little friends. Many restaurants are now providing such services during the New Year, and it has become a great business opportunity. People will even book the table weeks in advance.

Watching the Spring Festival Gala Evening. On every New Year Eve, CCTV will host the Spring Festival Evening, which includes by actors from different provinces and regions, including Hong Kong, Macao and Taiwan. It is regarded as an evening with the largest audience ratings, the longest showing time and the biggest number of actors in the world. Since the first show in 1983, more and more Chinese families have been enjoying and watching the event. They will do this while eating New Years Dinner with the family and having fun with the super stars on the screen, instead of sitting together to chat and listening to the older people talk about family history.

Development of New Year Movies. In the past, people used to stay at home, to play mahjong and to chat, and the streets would be empty. As living standards improve, the cultural requirements of Chinese people are being improved. Watching a New Year Movie has become an important part of cultural activities in China, especially for the younger generations. In order to satisfy the audience with fun, such movies are usually comedies and action films, with great a great storyline, beautiful scenery and excellent entertainment, to attract more and more people. One of the most famous directors of New Year Movies, is Mr. Feng Xiaogang. His famous works include "Dream Factory", "Be There or Be Square", with a huge market impact and high box office revenues.

Tours of the 7-day holiday. In the Chinese tradition, Spring Festival is the time of reunion. In the past, people were used to staying at home or visiting their friends or relatives. However, as Spring Festival became the longest public holiday, a whole week from January 1st to 7th, this period has become a golden week of tourism. Now many people are no longer staying at home, but taking the long holiday to travel to different places to spend the Spring Festival with their families. This way, people can not only get together, but also have more fun, staying away from their busy city lives and enjoying the ease of tourism. At the same time, people are traveling to different places to meet different kinds of people, to learn about different cultures and to experience different customs, are also important reason for many travelers.

No custom is formed in a short period of time. It is a long-term habit developed by the whole society. The continuous development of Chinese Spring Festival customs or the diminishing or changing of the old ways are just part of the iceberg. The spirit and core values of Chinese New Year still remain unchanged. That is the call for love, peace and happiness among fellow people. Actually, the spirit and core value of it is growing and adapting with the Chinese modern lifestyle. In a word, the color of Spring Festival never fades.

4. Spring Festival for Overseas Chinese

As China becomes a more important country in the modern day world and with the growth of Chinese influence of overseas, Spring Festival has been recognized and accepted by more and more different countries. It has also become the best carrier of Chinese culture in the study of China for the entire world.

Now, the Chinese Spring Festival has become a popular festival in London. Before every Chinese New Year, different Chinese Medias start to show different factual programs about the Chinese Spring Festival, which increase the curiosity about China among foreigners. People also start to study about their open zodiacs and the Zodiacal Animal of that year. During the Spring Festival, in China Town, all the families will hang lanterns to light up the whole street, and people will be surprised to find different kinds of New Year Goods including Tang suits, lanterns, strings of firecrackers, scrolls, paintings, Chinese knots and so on being sold everywhere in the supermarket. Such goods are very popular among people and the markets will become extremely crowded. On New Year's Eve, people will make dumplings with their Chinese friends and enjoy the Chinese New Year with everyone they meet. Many Chinese restaurants will serve different kinds of New Years dishes like the fish dish named "Affluent Year" and the chicken dish named "Phoenix Year". It's the taste of a traditional Chinese New Year for everyone who wants to take part.

The New Year Parade is one of the favorite events for tourists and locals alike. To increase the New Year atmosphere, different Chinese groups will organize the traditional shows on the streets, like the dragon dance, the lion dance and so on. Or they will visit different stores of the streets to give New Year greetings and wishes. People will also take part in parades, different kinds of operas, dances, songs, calligraphy, paper crafts and folk music. On New Years Eve, the sky of London will be lit by colorful fireworks. In the warm glow of the fireworks, everybody is welcome to enjoy the show and hopefully it will bring a smile to all their faces. No matter who you are and no matter if you are able to speak Chinese, people will greet each other in Chinese with sayings such as "Good Fortune in the New Year" or "Happy New Year".

During the Spring Festival, Trafalgar Square in central London is also full of cheer, with tens of thousands of people gathering to celebrate the Chinese New Year. The sounds of fireworks and drums will light up the passion of every visitor. The performances of the dragon dances and the lion dances also bring on the excitement of the New Year. Organized by the party host, the audience will chant in one voice sayings like "Good Fortune" and "How Are You", throughout the evening. Some people will wear the costume of the God of Fortune and give red envelopes to people. All kinds of different excellent and cultural shows combine to be a great feast of culture. Everybody is encouraged to enjoy the Chinese folk culture and the Chinese festival.

Different shows present different cultural attractions to people. So it has increased the friendship between Chinese people and most foreigners. Such activities are supported not only by many big companies in our country, but also by government authorities and organizations of China and Britain, like the local Tourism Bureaus, Banks and types of Media. The government has taken the opportunity to promote "Chinese Culture in London", to encourage people to learn the history of Chinese culture. People will hopefully also learn and gain confidence about the next year through the study of thee different Chinese Zodiacal Animals of the Spring Festival.

Now, encouraged by the government policies of diversified cultures, the Chinese New Year celebrations in London have attracted a large audience and are being participated in by more and more people. It has become the biggest Spring Festival event outside of Asia. So it has been regarded as one of the main local cultures in London. In each Chinese New Year, people enjoy the shows on Trafalgar Square, Leicester Square or China Town, like the dragon dance and the lion dance, and enjoy thee taste of the Chinese food. The Chinese New Year with fireworks and happy laughter has become an exciting new event for Londoners.

Not only in China and London, the Spring Festival celebrations in the US are also extremely popular. During the Spring Festival, colorful scrolls reading "Happy New Year" and "Happiness" are hung everywhere on the streets. Lucky pictures will also be placed in the windows of different shopping centers and department stores. With booths selling special Chinese New Years Goods will also be set up, where people can find different kinds of Chinese New Year commodities. In recent years, the Chinese Spring Festival culture has been accepted by more and more locals abroad, and companies like Disney also send cartoon characters to participate in the shows. The celebration activity of the Chinese New Year is also supported by the American government. Americans are also able to see the fireworks, one of the traditional activities of the Chinese Spring Festival. The government of New York has set up special places and times for people to enjoy the fireworks, including China Town, where Chinese people live together. Stamps of Chinese Zodiacal Animals will also be published by the American Post Offices, to memorize the New Year. And in 2003, New York brought in Spring Festival as a public holiday for people to enjoy. You can see the influence of Spring Festival in US more and more. Americans enjoy the Chinese cultures and regard Spring Festival as a perfect opportunity to learn about traditional Chinese cultures.

Now, although Spring Festival is not a public holiday in Canada, more and more people from the main society are paying attention to it. New Year Visits have been put as an important schedule for some government officials and company managers. Many of them are also speaking Chinese wishing others for "Good Fortune". In Spring Festival Parades, each year they take place in cities like Toronto and Vancouver, many important figures of local governments and high profile individuals of the media will participate in the activities and many of them will even wear the Tang Suits and send out red envelopes to people on the streets. The items stuffed into the envelope are always a big surprise. Sometimes there is money, but usually there are chocolates. Combined with the local custom and lifestyle, it is growing more popular every year.

5. Appendix

A. Query Table of the 12 Zodiac Animals and the relevant personality analysis of people born in different years

a. Query Table of the 12 Zodiac Animals

Zi	Chou	Yin	Mou	Chen	Si	Wu	Wei	Shen	You	Xu	Hai
Rat	Ox	Tiger	Hare	Dragon	Snake	Horse	Ram	Monkey	Rooster	Dog	Pig
1900	1901	1902	1903	1904	1905	1906	1907	1908	1909	1910	1911
1912	1913	1914	1915	1916	1917	1918	1919	1920	1921	1922	1923
1924	1925	1926	1927	1928	1929	1930	1931	1932	1933	1934	1935
1936	1937	1938	1939	1940	1941	1942	1943	1944	1945	1946	1947
1948	1949	1950	1951	1952	1953	1954	1955	1956	1957	1958	1959
1960	1961	1962	1963	1964	1965	1966	1967	1968	1969	1970	1971
1972	1973	1974	1975	1976	1977	1978	1979	1980	1981	1982	1983
1984	1985	1986	1987	1988	1989	1990	1991	1992	1993	1994	1995
1996	1997	1998	1999	2000	2001	2002	2003	2004	2005	2006	2007
2008	2009	2010	2011	2012	2013	2014	2015	2016	2017	2018	2019
2020	2021	2022	2023	2024	2025	2026	2027	2028	2029	2030	2031
2032	2033	2034	2035	2036	2037	2038	2039	2040	2041	2042	2043
2044	2045	2046	2047	2048	2049	2050	2051	2052	2053	2054	2055
2056	2057	2058	2059	2060	2061	2062	2063	2064	2065	2066	2067
2068	2069	2070	2071	2072	2073	2074	2075	2076	2077	2078	2079
2080	2081	2082	2083	2084	2085	2086	2087	2088	2089	2090	2091
2092	2093	2094	2095	2096	2097	2098	2099	2100	2101	2102	2103
Zi	Chou	Yin	Mou	Chen	Si	Wu	Wei	Shen	You	Xu	Hai
Rat	Ox	Tiger	Hare	Dragon	Snake	Horse	Ram	Monkey	Rooster	Dog	Pig

People born in the Year of Rat: optimistic, smart, humorous, skillful and popular. They are proactive, industrious, lucky for the whole of their life and capable to conquer occasional difficulties. With great viability, they can handle different circumstances skillfully. Though they are modest, they never act aggressively and are even sometimes considered unimportant, they keep working hard with consistency. And they are never easily defeated.

People born in the Year of Ox: practical, diligent, careful, realistic, responsible, serious in working attitude, appreciated and reliable to the manager in their work life and able to conquer difficulties with great endurance. Most of these people are conservative, shy and honest. They are rarely romantic but straight forward to express their happiness, anger, sadness and joy. They can never stand to deal with people that they don't like. But they can take good care of the people who are relying on them. Usually they make good friends with the ones they like. When they feel angry, they will act like a wild ox, which can't be easily calmed by anyone.

People born in the Year of Tiger: independent, extremely self-esteemed, acting alone, bossy and protective to the weak. They are steady and assertive in their career, no matter how hard the situation. They will never quit. However they are radical, reckless and easy to make mistakes. That's usually the reason why they will often lose at many different types of games. They will be assertive and constant facing difficulties and loss. On the other hand, they are aggressive and never feel defeated. They are straight-forward with great sense of justice. Even if they are arguing with their supervisors, they will stick to the truth.

People born in the Year of Hare: They appear to be quiet, soft and fragile. However, they are confident with a strong will. They can act on their plans and according to their goals. They are polite and gentle even to their enemies. They are skillful and quick to avoid any risks. And of course, when attacked, they will take appropriate actions to protect their own interests. They can skillfully jump over any difficulties and quickly recover from defeat. Believing in their own viability and relying on their own judgment, they can easily find happiness and satisfaction in their life.

People born in the Year of Dragon: mysterious as a dragon, they are elusive, ambitious, adventurous and romantic. At the same time, they are calm and always act with great self importance. They will be more active than others. Once they start to take actions, they will stick to their goals. They are generous, with vibrancy and strength. They are charming and bossy, with strong sense of objective. Once they make up their minds, no one can change or defeat them. They are tough and assertive. They are not tricky, but always confident in sticking to their dreams. Therefore, they sometimes are numb to the things happening around them and can find the right ways to solve the problems.

People born in the Year of Snake: mysterious, scholarly, romantic, calm and consistent. They are modest and sticking to set plans. They are quiet and never get easily annoyed. They are thoughtful and intelligent. They know their own capability and love mental satisfaction, with a strong sense of judgment. They are sharp in mind with intelligent thoughts. Though appearing quiet, they can make quick and assertive decisions. However, the people born in this year have a strong sense of possessiveness. This is the weak point of their personality. People feel it is difficult to make friends with them, and they never easily share their thoughts. They are also picky about friends. Being modest, however, they are stubborn and never feel defeated.

People born in the Year of Horse: Cheerful, energetic and quick-minded, with strong insight and mental independence, they often start their own business when they are still young. With strong confidence and a nice personality, they never stick to the set rules. They can react quickly and make the right decisions. But sometimes they appear to be radical, stubborn and hot-tempered. However, they are also forgetful. They have strong goals but they can't well recognize their own weakness. Neither can they change immediately. With bad implementation, especially when solving big problems, they are often satisfactory about the small achievements and forget about the next steps. They usually can't concentrate well.

People born in the Year of Ram: honest, kind, compassionate, gentle and even a little shy. They can easily make up for mistakes made by others and are considerate about their difficulties. No matter where they go, they like to communicate with other people and treat them with honesty. However, sometimes, the easy-going appearance and the outer assertiveness are conflicting. When arguing with others, they would rather to keep silent with anger, than trying to persuade or explain repeatedly. Neither would they show their disappointment. When they are scared, they react drastically. Sometimes they feel sad and sentimental. They look at things with pessimistic eyes. What they need is the passion and support from people around them.

People born in the Year of Monkey: being passionate, confident, reliable and smart, they can deal with different difficult situations calmly. They would rather not waste a minute when they are working. Though some of them appear to be shy, they actually would not change their minds easily. They tend to work hard for consistency, otherwise they will complain and refuse to work any more. They have strong sense of competition and they would rather hide their real thoughts. They will also make their own plans by themselves. With strong superiority, sometimes they don't show respect to others. They always stick to their own interests, gain or loss.

People born in the Year of Rooster: energetic, smart, well-organized, serious, honest, assertive, ambitious and successful. They are always straightforward and aggressive to criticize cruel behavior. They have a good reputation at work and they know how to obtain the trust from their managers with intelligence and high efficiency. When asked to do anything they are not willing to, they will not waste any time to even have an attempt at the task. Most of the people born in this year are picky and want everything to be perfectly set up. Controlled by the strong sense of competition, when annoyed, they will not hesitate to fight back with extreme measures.

People born in the Year of Dog: honest, reliable, fair, industrious and willing to be a good listener and to share sadness with their friends. They will never surrender the truth that they believe in. Once they make up their minds, they will not be influenced by any outer force. When arguing with others, they tend to defeat their opponents with perfect logic. When they feel disturbed or threatened, they will also take drastic measures to fight back. When fighting with others, they are usually open and straightforward, like a warrior.

People born in the Year of Pig: steady, industrious, kind-hearted, honest, warm-hearted and forgiving. Therefore, they can always keep good relationships with other people. Of course, when cornered, they will also fight back. However, they also hate conspiracy and jealousy. They stick to their goals and are reliable to complete all the work that has been tasked to them, with their best efforts. Though trying to be patient, they are easily annoyed. They could be peaceful and happy one minute and turn suddenly the next.

B. Development Chart of the Chinese characters of the 12 Zodiac Animals

鼠——Rat   牛——Ox   虎——Tiger   兔——Hare

龙——Dragon   蛇——Snake   马——Horse   羊——Ram

猴——Monkey   鸡——Rooster   狗——Dog   猪——Pig

Oracle-→Small Seal Style-→Regelschrift

C. Purchasing List of Spring Festival Goods

Ornaments: 1 big Chinese Knot, 3 sets of cloth peppers of mixed colors, 2 pieces of paper crafts of character for Luck, 2 window paper silhouettes, 2 spring festival couplets, 2 Door God pictures, 2 red lanterns, fresh flowers and so on

Drinks: Juice, wines, Chinese teas and so on

Snacks: All kinds of candy, chocolate, plum and melon seeds, peanuts, pistachios, almonds, red dates, dried persimmons, pine nuts, walnuts and other nuts

Fruits: Grapes, apples, grapefruit and so on.

Dishes: Fresh pork, fresh meat, fresh beef, fresh fish, fresh crab and shrimp, eggs, abalone, sea cucumber, fish maw, scallops and mushrooms, fungus and so on, preserved chicken, preserved duck, preserved sausage, rice pudding, and dishes of certain vegetables and secondary dishes.

Staple foods: Frozen dumplings, sticky rice balls, bread, sticky rice cakes, ciba and so on.

Garments: A set of Tang Suits (mainly red)

Games: 2 sets of firecrackers, 1 box of fireworks and 4 gun salutes

Others: 12 red envelopes and 12 New Year Cards and so on